BILLY CARSON'S PINEAL REVOLUTION:
Awakening Inner Vision

Oteren.Fredrick

DISCLAIMER

The data gave in this book is planned for instructive and enlightening purposes just and ought not be understood as clinical or mental exhortation. The substance depends on the creator's exploration, individual experiences, and accessible logical writing. In any case, it doesn't fill in for proficient clinical, mental, or remedial guidance.

COPYRIGHT

© [2024] [Oteren.Fredrick]. Protected by copyright law.

No piece of this book might be duplicated, conveyed, or communicated in any structure or using any and all means, including copying, recording, or other electronic or mechanical techniques, without the earlier composed consent of the distributer, with the exception of brief citations typified in basic audits and certain other noncommercial purposes allowed by intellectual property regulation.

TABLE OF CONTENT

INTRODUCTION
CHAPTER 1:
- The Pineal Gland: An Ancient Mystery
- Interfacing Old Understanding and Present day Science

CHAPTER 2:
- Billy Carson's Journey
- Building a Scaffold Among Science and Otherworldliness

CHAPTER 3:
- Understanding the Pineal Gland
- Life designs and Physiology of the Pineal Organ

CHAPTER 4:
- Activating the Pineal Gland
- Third Eye Meditation
- Care Practices
- Diet and Sustenance
- Decrease Fluoride Exposure
- Sunshine Exposure
- Rest Hygiene
- Detoxification

- Portrayal Techniques
- Affirmations

CHAPTER 5:
- The Pineal Gland and Spiritual Awakening
- The Pineal Organ in Out of date Traditions
- The Occupation of the Pineal Organ in Powerful Experiences
- Practices for Working with Significant Stimulating
- Breath of Fire
- The Om Mantra
- Decrease Electromagnetic Fields (EMFs)
- Customary Light Exposure

CHAPTER 6:
- The Science Behind the Mysticism
- The Development and Capacity of the Pineal Organ
- The Pineal Organ and Neural connections
- The Pineal Organ and Light Wisdom
- Decalcifying the Pineal Organ

CHAPTER 7:
- Practical Applications

- Reflection Techniques for Pineal Organ Order
- Insight Techniques
- Dietary Practices for Pineal Organ Prosperity
- Lifestyle Changes for Further developing Pineal Organ Capacity
- Involving Development for Pineal Organ Inception

CHAPTER 8:
- Challenges and Misconceptions
- Normal Difficulties in Pineal Organ Enactment
- Tending to Difficulties and Confusions Practically speaking

CONCLUSION:
APPENDIX
ACKNOWLEDGEMENTS

INTRODUCTION

The human experience is a many-sided dance of psyche, body, and soul. At the crossing point of these aspects lies a little, puzzling organ that has interested researchers, thinkers, and otherworldly searchers for quite a long time: the pineal organ. Frequently alluded to as the "third eye," this little endocrine organ, situated in the focal point of the mind, has been respected in different societies and profound customs for its alleged association with higher conditions of cognizance and otherworldly edification.

In the domain of current profound investigation, scarcely any figures stand out and regard as Billy Carson. An acclaimed specialist, creator, and speaker, Carson has committed his life to uncovering the privileged insights of old information and its applications to contemporary profound practices. His work on the pineal organ has ignited an upset by they way we figure out this mysterious piece of our life systems, overcoming any issues between old insight and current science.

This book, "Billy Carson's Pineal Upset: Arousing Internal Vision," is an excursion into the profundities of Carson's experiences and disclosures with respect to the pineal organ. We will investigate the verifiable and social meaning of the pineal organ, dive into Carson's own excursion and arousing, and reveal the logical reason for the profound elements of this uncommon organ.

Through this investigation, you will find down to earth strategies and practices prescribed by Billy Carson to initiate and sustain your pineal organ, opening new elements of mindfulness, imagination, and otherworldly development. We will likewise address normal confusions and difficulties related with pineal organ actuation, giving you the information and instruments to set out on your own excursion toward inward vision and edification.

Plan to set out on an extraordinary journey that rises above the physical and dives into the domains of the otherworldly. Whether you are a carefully prepared profound searcher or new to the idea of the pineal organ, this book plans to give you an extensive comprehension and a functional manual for arousing your inward vision.

Welcome to the unrest. Welcome to the enlivening. Welcome to the excursion of finding the force of your pineal organ through the significant experiences of Billy Carson.

CHAPTER 1:

The Pineal Gland: An Ancient Mystery

The pineal organ, a little endocrine organ got comfortable the point of convergence of the brain, has been a subject of interest and mystery for quite a long time. Regarded in various obsolete social orders and supernatural practices, it is habitually suggested as the "third eye" in view of its clear work in higher states of mindfulness and significant information. This segment researches the valid, social, and consistent perspectives on the pineal organ, following its frustrating outing from old-fashioned times to momentum understandings.

Bona fide Perspectives
The pineal organ's significance follows as far as possible back to old-fashioned human progressions, where it was regularly associated with significant exciting and enlightenment. The old Egyptians, for instance, loved the pineal organ as the seat of the soul. They depicted it symbolically in their strength and plan, most surprisingly as the Eye of Horus. This picture, tending to security, renowned power, and extraordinary prosperity, appears to be like the

actual development of the pineal organ, highlighting its loved status in Egyptian supernatural quality.

In old India, the possibility of the third eye, or "Ajna Chakra," is a central standard in the demonstration of yoga and reflection. Arranged between the eyebrows, this energy local area is acknowledged to be the entryway to higher perception and significant comprehension. The demonstration of fortifying the third eye through consideration and different yogic techniques expects to mix the pineal organ, working with a more significant relationship with the magnificent.

Basically, in old China, the pineal organ was connected with the "Upper Dantian," one of the three fundamental energy networks in the body. Taoist practices underline the advancement of internal energy, or "Qi," through examination and hand to hand battling, completely plan on starting the Upper Dantian and as such overhauling significant care and life expectancy.

The Pineal Organ in Western Thought

In Western thought, the pineal organ's significance was brought to perceptible quality by the French mastermind and mathematician René Descartes in the seventeenth 100 years. Descartes comprehensively implied the pineal organ as the "seat of the soul," proposing that it was the sign of relationship between the cerebrum and the body. He acknowledged that the pineal organ expected a dire part in mediating between the physical and the powerful spaces, a thought that has influenced following philosophical and legitimate examinations concerning the possibility of discernment.

Regardless of Descartes' philosophical statements, laid out scientists stayed by and large dubious of the pineal organ's suggested powerful abilities. For an enormous piece of the nineteenth and mid 20th many years, the pineal organ was seen as an insignificant organ with no colossal capacity, like the reference segment. This acumen began to change during the 20th hundred years with the revelation of the compound melatonin, conveyed by the pineal organ.

Legitimate Disclosures
The disclosure of melatonin during the 1950s changed the legitimate perception of the pineal organ. Melatonin is a substance that oversees rest wake cycles and circadian rhythms, interfacing the pineal organ to essential physiological cycles. Research has shown that the improvement of melatonin is affected by light transparency, with additional raised levels conveyed in cloudiness, supporting the chance of the pineal organ as a "third eye" that answers natural light signals.

Further consistent assessments have uncovered that the pineal organ contains photoreceptor cells like those found in the eyes, suggesting that it could have created from a rough light-sensitive organ. This finding maintains old significant feelings that the pineal organ is a section to higher knowledge and care.

Despite melatonin, the pineal organ has been found to make follow proportions of other biochemicals, for instance, serotonin, a neurotransmitter related with disposition rule and impressions of success. The presence of these substances in the pineal organ has driven a couple of experts to suggest that it could expect a section in adjusting up close and personal and mental states, further conquering any issues among physical and significant prosperity.

The Pineal Organ in Contemporary Power
The resurgence of interest in the pineal organ in contemporary power can be credited to the transformation of old knowledge and current coherent revelations. Extraordinary searchers and experts of various disciplines, including examination, yoga, and far reaching prosperity, view the pineal organ as a critical part in achieving higher states of comprehension and significant enlightenment.

Current extraordinary teachers, similar to Billy Carson, highlight the meaning of activating the pineal organ through various strategies, including thought, care, and unequivocal dietary and lifestyle practices. These procedures are acknowledged to decalcify the pineal organ, which can become calcified as a result of components like not exactly heavenly eating schedule, fluoride receptiveness, and nonappearance of light. Decalcification is made sure to update the pineal organ's ability, working with additional significant extraordinary experiences and pieces of information.

Billy Carson, explicitly, has transformed into an obvious figure in the contemporary examination of the pineal organ. Bringing from his wide assessment into old messages, consecrated math, and present day science, Carson gives an exhaustive construction to getting a handle on the pineal organ's work in significant stirring. He sets that the establishment of the pineal organ can provoke huge changes in wisdom, creative mind, and mindfulness, allowing individuals to exploit their internal vision and open their most extreme limit.

Interfacing Old Understanding and Present day Science

The combination of old knowledge and current science offers an exhaustive perspective on the pineal organ, seeing its different work in human physiology and extraordinary quality. Old social orders intuited the significance of the pineal organ through their powerful practices and meaningful depictions, while present day science gives exploratory evidence of its physiological abilities and potential for further developing success.

This exhaustive cognizance invites us to explore the pineal organ not just as a characteristic organ yet rather as a crucial perspective for opening higher parts of care and significant turn of events. It urges us to embrace practices that support the pineal organ, empowering a more significant relationship with ourselves and the universe.

The pineal organ stays one of the most enrapturing and cryptic pieces of the human body. Its significance ranges across social orders and ages, reflecting humankind's helping through mission for powerful enlightenment and higher awareness. From the old Egyptians and Hindus to introduce day supernatural searchers and analysts, the pineal organ continues to fascinate our inventive psyche and move our mission for internal vision.

As we plunge further into the encounters and practices recommended by Billy Carson, we will uncover the historic ability of the pineal organ and how it can modify how we could decipher insight and power. In the going with segments, we will explore practical procedures to authorize and support the pineal organ, traversing the old knowledge with contemporary coherent disclosures to mix our internal vision and set out on a trip of critical significant stimulating.

CHAPTER 2:

Billy Carson's Journey

Billy Carson, an acclaimed scientist, creator, and profound educator, has dazzled crowds overall with his profound bits of knowledge into old insight and present day science. His excursion to turning into a main voice on the pineal organ and otherworldly arousing is one set apart by interest, persistent quest for information, and significant individual encounters. This part dives into Carson's experience, the key impacts that molded his profound way, and the significant minutes that prompted his profound comprehension of the pineal organ.

Early Life and Arousing
Billy Carson's process started in the metropolitan scene of New York City, where he was brought up. In spite of the difficulties of experiencing

childhood in a clamoring city, Carson showed an early interest in science, history, and the secrets of the universe. His curious nature drove him to address regular stories and investigate elective viewpoints on life and presence.

Carson's underlying introduction to the universe of old human advancements and elusive information got through his interest with Egyptology. The glory of the pyramids, the complexities of hieroglyphics, and the significant insight of the old Egyptians started a long lasting enthusiasm for uncovering stowed away bits of insight. This early interest set up for his later work in disentangling the profound and logical meaning of old curios and messages.

Finding the Pineal Organ
Carson's advantage in the pineal organ was aroused during his broad investigations of old societies. He saw repeating references to a magical "third eye" across different civic establishments, from the Eye of Horus in Egypt to the Ajna Chakra in India. These images

highlighted an inward vision and higher condition of cognizance that fascinated Carson.

His exploration drove him to crafted by René Descartes, the French rationalist who broadly named the pineal organ the "seat of the spirit." Descartes' affirmation that the pineal organ was the mark of association between the brain and the body resounded with Carson's journey to figure out the transaction among otherworldliness and science. This convergence turned into a point of convergence of his examinations, as he looked to connect old insight with contemporary logical bits of knowledge.

Key Impacts and Coaches
All through his excursion, Carson experienced a few vital impacts and tutors who assumed a pivotal part in forming how he might interpret the pineal organ and otherworldly arousing. Among them was Drunvalo Melchizedek, a famous creator and educator known for his work

on consecrated calculation and cognizance. Melchizedek's lessons on the MerKaBa contemplation and the initiation of the light body gave Carson reasonable devices for his otherworldly practice.

Another huge impact was Gregg Braden, a researcher and creator who investigates the intermingling of science and otherworldliness. Braden's work on the force of the heart, the field of awareness, and the human potential for self-mending resounded profoundly with Carson's journey to grasp the full abilities of the human brain and body.

Individual Encounters and Leap forwards
Carson's process was not exclusively educated; it was likewise profoundly private and experiential. He took part in different reflection works on, including those pointed toward enacting the pineal organ and arousing the third eye. These practices prompted significant

changes in his discernment and awareness, confirming the lessons he had contemplated.

One vital experience happened during a reflection retreat where Carson rehearsed delayed haziness contemplation, a procedure utilized in numerous otherworldly practices to invigorate the pineal organ. Without light, the pineal organ's creation of melatonin increments, possibly prompting the discharge of other biochemicals, for example, dimethyltryptamine (DMT), which is accepted to work with enchanted encounters.

During this retreat, Carson revealed encountering clear dreams, improved instinct, and a significant feeling of interconnectedness with the universe. These encounters hardened his faith in the groundbreaking force of the pineal organ and built up his obligation to imparting this information to other people.

Building a Scaffold Among Science and Otherworldliness

Billy Carson's extraordinary commitment lies in his capacity to blend old insight and current science, introducing a comprehensive comprehension of the pineal organ and its true capacity for otherworldly arousing. He stresses that the pineal organ isn't only a mysterious idea yet an unmistakable piece of our physiology that can be sustained and initiated through unambiguous practices.

Carson's work includes broad examination into both old texts and contemporary logical investigations. He digs into the compositions of antiquated human advancements, deciphering images and similitudes that highlight the meaning of the pineal organ. All the while, he stays informed concerning the most recent logical exploration on neurobiology, endocrinology, and awareness, utilizing this information to approve and develop antiquated lessons.

Instructing and Effort

Perceiving the groundbreaking capability of his disclosures, Billy Carson has devoted his life to instructing and exceed. He established 4biddenknowledge Inc., a stage that gives instructive substance on old history, otherworldliness, and the secrets of the universe. Through books, recordings, and courses, Carson imparts his bits of knowledge to a worldwide crowd, rousing people to investigate their own otherworldly potential.

His book, "Summary of the Emerald Tablets," is a demonstration of his obligation to spanning old insight and present day understanding. In this work, Carson translates the lessons of Thoth, the Atlantean, giving perusers down to earth direction on enacting the pineal organ and accomplishing higher conditions of cognizance.

Carson's effort stretches out past conventional stages. He effectively draws in with his crowd through online entertainment, online courses, and live occasions, making a local area of similar people devoted to individual and aggregate arousing. His receptive style and certified enthusiasm for his topic make his lessons open to a wide crowd, from prepared otherworldly searchers to those new to the way.

Billy Carson's process is a demonstration of the force of interest, determination, and the mission for more profound comprehension. His one of a kind capacity to connect old insight and current science has enlightened the meaning of the pineal organ in otherworldly arousing and individual change. Through his exploration, individual encounters, and lessons, Carson gives a guide to people trying to open their internal vision and tap into their maximum capacity.

As we keep on investigating the experiences and practices suggested by Billy Carson, we will reveal functional procedures for enacting the pineal organ and coordinating its groundbreaking power into our regular routines. The excursion toward profound arousing is both an individual and aggregate undertaking, and

Carson's work fills in as a directing light on this way.

CHAPTER 3:

Understanding the Pineal Gland

The pineal organ, habitually suggested as the "third eye," is a little endocrine organ organized in the point of convergence of the psyche. No matter what its moment size, the pineal organ has enchanted the thought of both old mystics and present day specialists due to its obvious work in associating the physical and significant areas. In this part, we will research the existence frameworks and physiology of the pineal organ, its work in the human body, and its relationship with perception.

Life designs and Physiology of the Pineal Organ

The pineal organ is formed like a little pine cone, in this manner its name, and is by and large the size of a grain of rice. It is arranged near the point of convergence of the frontal cortex, between the different sides of the equator, got comfortable an indent where the two pieces of the thalamus join. Despite its little size, the pineal organ is sumptuously furnished with blood, getting a higher stream for every unit volume than another organ, beside the kidneys.

The pineal organ is basically made from pinealocytes, the phones at risk for conveying melatonin. Likewise, it contains glial cells, which proposition help and protection for the pinealocytes. The organ is enveloped by a holder of connective tissue and is habitually surrounded by calcium stores, which can end up being more unavoidable with age, an eccentricity known as calcification.

Melatonin: The Rest Regulator

The most eminent capacity of the pineal organ is the production of melatonin, a synthetic that controls rest wake cycles and circadian rhythms. Melatonin association is influenced by the light-faint cycle; its creation developments on account of lack of definition and reduces with light receptiveness. This beat is obliged by signals from the suprachiasmatic center (SCN) of the operational hub, which gets input from the eyes about the external light environment.

Melatonin has an extent of ramifications for the body past coordinating rest. It goes probably as a malignant growth counteraction specialist, protecting cells from hurt achieved by free fanatics. It in like manner influences conceptive synthetic compounds, expecting a section in the preparation of pubescence and periodic recreating in animals. Furthermore, melatonin has been shown to control safe capacity, exhibiting its greater significance in staying aware of in everyday prosperity.

The Pineal Organ and Light Insight

One of the most intriguing pieces of the pineal organ is its antipathy for light. The organ contains photoreceptor cells like those found in the retina of the eyes, which suggests that it could have created from an unrefined light-delicate organ. This has driven a couple of researchers to suggest that the pineal organ is a negligible "third eye" that once expected a more clear part in light wisdom.

In lower vertebrates, for instance, reptiles and animals of land and water, the pineal organ is point of fact arranged to do clearly perceiving light and affecting behavior similarly. For example, the parietal eye of some reptile species is a utilitarian photoreceptive organ related with the pineal organ, helping with controlling circadian rhythms and intermittent approaches to acting.

The Pineal Organ and Serotonin

Despite melatonin, the pineal organ conveys and cycles serotonin, a neural connection that expects a basic part in mentality rule, hankering, and rest. Serotonin is every now and again implied as the "vibe extraordinary" neural connection on account of its effect on success and euphoria.

The association among serotonin and melatonin is particularly intriguing. Serotonin fills in as a precursor to melatonin, and its levels in the pineal organ change in view of the light-dull cycle. During the day, when light receptiveness is high, serotonin levels are raised. Around night time, the impetus serotonin N-acetyltransferase (NAT) changes serotonin over totally to N-acetylserotonin, which is then moreover exchanged over totally to melatonin. This change is a basic piece of the communication that controls rest wake cycles.

The Pineal Organ and Mindfulness
The pineal organ's relationship with mindfulness is one of the most interesting and examined pieces of its capacity. Old significant traditions and present day experts the equivalent have

proposed that the pineal organ is a way to higher states of comprehension and supernatural comprehension.

One of the biochemicals related with the pineal organ's work in perception is dimethyltryptamine (DMT), a strong stimulating compound. DMT is essentially similar to serotonin and melatonin and is typically made in restricted amounts in the human body. A couple of experts acknowledge that the pineal organ could make DMT, particularly during times of raised comprehension, for instance, brushes with death, significant reflection, or dreams.

DMT's ability to provoke huge changes in wisdom and perception has provoked its moniker as the "soul molecule." While the particular occupation of DMT in the pineal organ and its effect on comprehension remains a subject of ceaseless investigation, its presence maintains the possibility that the pineal organ could expect a section in working with extraordinary and otherworldly experiences.

Pineal Organ Calcification
One of the gigantic concerns with respect to the pineal organ's ability is calcification, the assortment of calcium stores inside the organ. Calcification can handicap the organ's ability to make melatonin and other biochemicals, conceivably impacting rest models and in everyday prosperity.

A couple of factors add to pineal organ calcification, including developing, diet, and regular receptiveness to fluoride. Studies have shown that general populations with high fluoride transparency will for the most part have higher speeds of pineal organ calcification. This has incited recommendations to lessen fluoride confirmation and support the pineal organ's prosperity through dietary and lifestyle choices.

Decalcifying the Pineal Organ
Decalcifying the pineal organ is a thought progressed by various supernatural specialists and experts who acknowledge that reducing

calcification can work on the organ's capacity and work with significant stimulating. A couple of frameworks are recommended to help the pineal organ and decrease calcification:

1. Diet: Gobbling up an eating routine rich in regular results of the dirt, close by food assortments high in cell fortifications, can maintain for the most part prosperity and reduce calcification. Avoiding dealt with food assortments, sugar, and fluoride-containing things is in like manner empowered.

2. Sunlight Exposure: Standard receptiveness to normal light can help with controlling the pineal organ's formation of melatonin and sponsorship circadian rhythms. Morning sunshine is particularly significant for setting the body's internal clock.

3. Meditation and Mindfulness: Practices, for instance, thought and mind can empower the pineal organ and advance loosening up and flourishing. Unequivocal techniques, such as

focusing in on the third eye during reflection, are acknowledged to activate the pineal organ.

4. Supplements: Certain upgrades, similar to iodine, magnesium, and boron, may maintain the pineal organ and decrease calcification. Also, flavors like turmeric, spirulina, and chlorella are known for their detoxifying properties.

The pineal organ, with its intricate plan and massive impact on both physiological and significant pieces of human life, remains a subject of mind boggling interest and mystery. Its part in coordinating rest, personality, and potentially perception features its significance in our overall success. As we dive further into the practices and encounters proposed by Billy Carson, we will continue to research how supporting the pineal organ can open new parts of care and significant turn of events.

Sorting out the pineal organ's life designs, works, and its relationship with higher states of awareness gives a foundation to the rational methodologies that will be inspected in coming about segments. These encounters range the old knowledge with present day consistent disclosures, guiding us on an outing towards stimulating our inner vision and achieving critical supernatural change.

CHAPTER 4:

Activating the Pineal Gland

The pineal organ, every now and again revered as the "third eye," holds the likelihood to open huge significant pieces of information and raised states of awareness. Sanctioning this otherworldly organ incorporates a mix of old practices and present day strategies highlighted working on its capacity and decreasing calcification. In this part, we will examine various strategies proposed by Billy Carson and other significant teachers for ordering the pineal organ, including reflection, diet, lifestyle changes, and normal factors.

Consideration and Care
Thought and care practices are among the best methods for enlivening the pineal organ. These practices help with quieting the mind, decrease tension, and advance a state of internal congruity, which are useful for starting the pineal organ.

Third Eye Meditation

Third eye reflection revolves around the district between the eyebrows, the region related with the pineal organ and the Ajna Chakra. This preparing remembers centering for this direct during examination toward vivify and mix the organ.

Directions to Practice Third Eye Meditation:
1. Find a Quiet Space: Pick a serene, pleasing place where you won't be vexed.

2. Sit Comfortably: Sit in a pleasing circumstance with your spine straight and eyes shut.
3. Focus on Your Breath: Take significant, slow breaths to relax your body and cerebrum.
4. Visualize the Third Eye: Guide your concentration toward the locale between your eyebrows. Imagine a splendid light or a sparkling circle here.
5. Chant or Repeat a Mantra: You can recount "Om" or any mantra that influences you to help with staying aware of focus.
6. Stay Present: In case your mind wanders, carefully return your thought to the third eye.
7. Practice Regularly: Hope to practice this examination ordinarily for something like 10-15 minutes.

Care Practices

Care incorporates staying present and totally participated in the continuous second. Standard consideration practice can reduce strain and advance a sensation of internal calm, which helps the pineal organ.

The best technique to Practice Mindfulness:
1. Focus on Your Breath: Spotlight on the energy of your breath as it enters and leaves your body.
2. Engage Your Senses: Notice the sights, sounds, aromas, and surfaces around you.
3. Observe Your Thoughts: Watch your contemplations without judgment and carefully return your focus to the ongoing second if your mind wanders.
4. Practice Daily: Coordinate consideration into your everyday regular practice, whether during blowouts, walks, or various activities.

Diet and Sustenance

The food you eat expects a basic part in the prosperity and capacity of your pineal organ. Certain food assortments and improvements can maintain the organ's ability and help with decreasing calcification.

Food assortments to Help Pineal Organ Health:
1. Raw Cacao: Affluent in malignant growth anticipation specialists and magnesium, unrefined cacao maintains in everyday brain prosperity.
2. Chlorophyll-rich Foods: Food assortments like spirulina, chlorella, and wheatgrass help with detoxifying the body and sponsorship pineal organ prosperity.
3. Iodine-rich Foods: Kelp and iodine improvements can help with decalcifying the pineal organ.
4. Turmeric: This zing has quieting properties and supports by and large frontal cortex prosperity.
5. Apple Juice Vinegar: Detoxifies the body and may uphold decalcification.

Supplements for Pineal Organ Activation:
1. Iodine: Supports the thyroid and may help with decreasing calcification.

2. Boron: Detoxifies the body and moving in everyday prosperity.
3. Magnesium: Principal for the larger part actual cycles and supports loosening up and rest.
4. Melatonin: Improving with melatonin can maintain the ordinary ability of the pineal organ.

Lifestyle Changes

Certain lifestyle changes can basically impact the prosperity and ability of your pineal organ. Incorporating these movements can update your overall thriving and backing your significant journey.

Decrease Fluoride Exposure

Fluoride is a commonplace ally of pineal organ calcification. Decreasing receptiveness to fluoride can help with staying aware of the sufficiency of your pineal organ.

Tips to Reduce Fluoride Exposure:

1. Drink Isolated Water: Use a water channel that dispenses with fluoride.
2. Choose without fluoride Toothpaste: Pick customary toothpaste without fluoride.
3. Avoid Dealt with Foods: Took care of food sources habitually contain fluoride and other disastrous engineered compounds.

Sunshine Exposure

Ordinary sunlight expects a vital part in controlling the pineal organ's improvement of melatonin and supporting overall circadian rhythms. Standard receptiveness to light, especially around the start of the day, can help your pineal organ.

Tips for Sunlight Exposure:

1. Spend Time Outdoors: Plan to spend something like 15-30 minutes outside each day.

2. Morning Sunlight: Morning sunshine is particularly beneficial for controlling your circadian rhythm.
3. Limit Counterfeit Light: Reduction receptiveness to fake light, especially blue light from screens, around evening time.

Rest Hygiene

Incredible rest neatness maintains the customary capacity of the pineal organ and the production of melatonin. Ensuring a solid rest plan and laying out a peaceful rest environment can update your overall success.

Tips for Better Rest Hygiene:

1. Maintain an Anticipated Rest Schedule: Make a beeline for rest and stir all the while reliably.
2. Create a Tranquil Environment: Make your room supportive for set down with open to bedding and unimportant light and upheaval.
3. Limit Stimulants: Avoid caffeine and electronic contraptions before rest time.

Detoxification

Detoxifying your body can maintain the capacity of the pineal organ and diminish calcification. Normal detox practices can update your overall prosperity and supernatural success.

Detoxification Practices:
1. Hydration: Hydrate to help with flushing harms from your body.
2. Sauna Therapy: Sweating in a sauna can help with taking out harms through the skin.
3. Epsom Salt Baths: Engrossing Epsom salt showers can assist with detoxification and loosening up.
4. Herbal Cleanses: Flavors like milk thistle, dandelion root, and burdock root can maintain liver detoxification.

Significant Practices
Despite genuine chips away at, participating in extraordinary practices can maintain the

activation of the pineal organ and work on your overall significant trip.

Portrayal Techniques

Portrayal strategies incorporate using your inventive brain to make mental pictures that help your targets and objectives. Envisioning the inception of your pineal organ can further develop your consideration practice.

The best strategy to Practice Visualization:
1. Find a Quiet Space: Sit or rests in a pleasing position.
2. Focus on Your Breath: Take significant, slow breaths to relax your body and cerebrum.
3. Visualize the Third Eye: Imagine a breathtaking light or a shining circle at the region of your third eye.
4. Feel the Activation: Envision the light growing further and feel the energy in your third eye area.

5. Practice Regularly: Coordinate discernment into your ordinary examination practice.

Affirmations

Accreditations are positive clarifications that can help with moving your viewpoint and sponsorship your points. Using affirmations associated with the pineal organ and significant stimulating can develop your preparation.

Occurrences of Affirmations:

1. "My pineal organ is sound and active."
2. "I am free to higher states of consciousness."
3. "I trust my internal vision and intuition."
4. "I am related with the eminent knowledge inside me."

Sanctioning the pineal organ is a perplexing trip that incorporates a blend of consideration, diet, lifestyle changes, and supernatural practices. By incorporating these methodologies into your regular day to day practice, you can maintain the prosperity and capacity of your pineal organ, working with additional significant powerful encounters and raised states of mindfulness.

Billy Carson's pieces of information and recommendations give a broad construction to starting the pineal organ, blending old understanding in with current science. As you continue to examine these practices, you will open new parts of care and set out on a remarkable journey toward powerful exciting.

In the going with areas, we will dive further into the logical uses of pineal organ establishment and research how to arrange these practices into your everyday daily schedule to redesign your overall thriving and extraordinary turn of events.

CHAPTER 5:

The Pineal Gland and Spiritual Awakening

The pineal organ has for a long while been connected with significant exciting and higher states of comprehension. Various old social orders and extraordinary practices consider it to be a section to interior vision and enlightenment. In this part, we will research the relationship between the pineal organ and powerful stimulating, the work it plays in otherworldly experiences, and the practices that can work with this tremendous change.

The Pineal Organ in Out of date Traditions

Since before time began, the pineal organ has been adored in various powerful and severe practices. It has been addressed as the "third eye" or the "seat of the soul" and is every now and again associated with illumination and higher circumstances.

Egyptian Culture
In old Egypt, the Eye of Horus addressed security, prosperity, and famous power, and it is acknowledged to address the pineal organ. The Egyptians pondered the Eye of Horus as a solid image of significant information and grand wisdom.

Hindu Tradition
In Hinduism, the pineal organ is connected with the Ajna Chakra, or the "third eye chakra," arranged between the eyebrows. The Ajna Chakra is seen as the point of convergence of sense, understanding, and spiritualist limits. Rehearses like reflection, yoga, and the use of blessed mantras plan to sanction this chakra, provoking significant stirring.

Buddhist Tradition

Buddhist illustrations complement the improvement of care and consideration to achieve enlightenment. The pineal organ is made sure to expect a section in achieving states of significant reflection and significance, helping experts with connecting with their genuine embodiment and general comprehension.

Christian Mysticism
In Christian powerful quality, the pineal organ is habitually implied as the "internal eye" or the "eye of the heart." A couple of understandings of scriptural texts recommend that the pineal organ is the "single eye" referred to by Jesus in the Uplifting news of Matthew, which, when established, fills the body with light and eminent knowledge.

The Occupation of the Pineal Organ in Powerful Experiences

The pineal organ's capacity to work with extraordinary experiences has been a subject of interest for both significant searchers and scientists. These experiences habitually incorporate a huge sensation of interconnectedness, expanded care, and a significant sensation of concordance and fortitude.

Dimethyltryptamine (DMT)
One of the key biochemicals related with the pineal organ's part in extraordinary experiences is dimethyltryptamine (DMT). DMT is major areas of strength for a compound regularly made in the human body. A couple of researchers recommend that the pineal organ integrates DMT, particularly during times of extreme tension, brushes with death, or significant consideration.

DMT's effects consolidate striking dreams, changed impression of presence, and a sensation of solidarity with the universe. These experiences line up with the portrayals of powerful feelings of excitement and captivated states found in various traditions.

Melatonin and Serotonin

Melatonin and serotonin, both made by the pineal organ, moreover expect critical parts in coordinating disposition, rest, and perception. Melatonin maintains significant, peaceful rest and is related with the rule of circadian rhythms. Serotonin, every now and again suggested as the "vibe incredible" neural connection, adds to impressions of success and happiness.

The exchange between melatonin, serotonin, and possibly DMT suggests that the pineal organ is complicatedly drawn in with making the conditions for extraordinary experiences and adjusted states of comprehension.

Practices for Working with Significant Stimulating

A couple of practices can help with starting the pineal organ and work with significant stimulating. These procedures intend to redesign inner vision, broaden comprehension, and partner with higher circumstances.

Examination and Breathwork

Examination and breathwork are fundamental practices for starting the pineal organ and empowering significant stirring. These strategies help with quieting the mind, lessen tension, and advance a state of inside concordance.

Breath of Fire

The Breath of Fire, a quick and cadenced breathing technique, is used in Kundalini yoga to enliven the pineal organ and sanction the third eye.

Guidelines to Practice Breath of Fire:
1. Sit Comfortably: Find a pleasing arranged position with your spine straight.
2. Inhale Deeply: Take a full breath in through your nose.
3. Exhale Rapidly: Inhale out quickly through your nose while pulling your navel in towards your spine.
4. Continue Rhythmically: Continue with this quick, melodic unwinding for 1-3 minutes.

Portrayal Techniques

Portrayal methodology incorporate making mental pictures that help your significant points. Envisioning the commencement of your pineal organ can further develop your appearance practice and work with supernatural stirring.

Sanctified Geometry

Consecrated estimation incorporates the examination of shapes and models that are acknowledged to reflect the focal principles of the universe. Imagining sacred numerical shapes, similar to the Blossom of Life or the Merkaba, can help with authorizing the pineal organ and develop insight.

Discussing and Mantras

Discussing and the usage of mantras can invigorate the pineal organ and advance significant exciting. The vibration of unequivocal sounds can redesign reflection and work with additional significant states of mindfulness.

The Om Mantra

Discussing "Om," pondered the beginning phase sound of the universe, is acknowledged to resonate with the pineal organ and the third eye chakra.

Bit by bit guidelines to Work on Discussing Om:
1. Sit Comfortably: Find a pleasant arranged position with your spine straight.
2. Take a Significant Breath: Take in significantly through your nose.
3. Chant Om: As you inhale out, serenade "Om" step by step, feeling the vibration in your body.
4. Repeat: Continue discussing for 5-10 minutes, focusing in on the sound and vibration.

Regular Factors
The environment you live in can basically influence the prosperity and ability of your pineal organ. Laying out a consistent

environment can work on your significant practices and work with stirring.

Decrease Electromagnetic Fields (EMFs)

Electromagnetic fields (EMFs) from electronic devices can upset the pineal organ's ability and melatonin creation. Restricting receptiveness to EMFs can maintain the organ's prosperity.

Ways of lessening EMF Exposure:

1. Limit Screen Time: Lessen the usage of electronic devices, especially before rest time.
2. Use EMF Shields: Consider including EMF safeguarding devices for your phone and PC.
3. Create a sans emf Rest Environment: Keep electronic devices out of your room and temperament executioner Wi-Fi around night time.

Customary Light Exposure

Customary sunlight maintains the pineal organ's advancement of melatonin and controls circadian rhythms. Standard receptiveness to ordinary light can overhaul your overall flourishing.

Tips for Sunlight Exposure:

1. Spend Time Outdoors: Mean to spend something like 15-30 minutes outside each day.
2. Morning Sunlight: Morning light is particularly valuable for dealing with your circadian beat.
3. Limit Counterfeit Light: Reduce receptiveness to fake light, especially blue light from screens, around evening time.

Integrating Extraordinary Practices into Everyday presence

Integrating powerful practices into your everyday timetable can make solid areas for a for significant exciting and the commencement of the pineal organ.

Regular Routine

Spreading out a solid everyday timetable that integrates examination, care, and other significant practices can maintain your trip towards significant stirring.

Cautious Moments

Incorporate cautious minutes throughout the span of your day to remain related with your significant practice. This can consolidate short reflection gatherings, significant breathing, or fundamentally stopping briefly to see the worth in the ongoing second.

Neighborhood Support

Attracting with a neighborhood comparable individuals can offer assistance and inspiration on your significant trip. Participate in reflection social events, significant retreats, or online conversations to connect with others who share your tendencies.

The pineal organ's relationship with significant exciting and higher states of mindfulness is deeply grounded in old traditions and maintained by present day sensible pieces of information. By participating in rehearses like reflection, breathwork, discernment, and care, you can impel the pineal organ and work with critical significant change.

Billy Carson's illustrations offer an intensive method for managing understanding and starting the pineal organ, blending old knowledge in with contemporary data. As you continue to explore these practices and integrate them into your everyday presence, you will leave on a journey of self-revelation, interior vision, and powerful stirring.

In the going with parts, we will jump into valuable usages of pineal organ activation, researching how to saddle its power for mindfulness, recovering, and change. Through these practices, you will open new components of care and connection point with the more significant pieces of your being.

CHAPTER 6:

The Science Behind the Mysticism

The pineal organ's relationship with supernatural experiences and higher states of insight has fascinated the two mystics and scientists for quite a while. While old social orders regarded it as the seat of the soul and an entry to higher spaces, momentum science hopes to sort out the physiological and biochemical instruments fundamental its heavenly properties. In this part, we will research the sensible pieces of information into the pineal organ, focusing in on its development, ability, and the biochemical cycles that could associate it to powerful experiences.

The Development and Capacity of the Pineal Organ

The pineal organ is a bit, pinecone-shaped endocrine organ arranged in the point of convergence of the frontal cortex. No matter what its little size, it expects an essential part in overseeing different physiological cycles, fundamentally through the discharge of melatonin.

Life designs of the Pineal Gland

- Location: The pineal organ is organized near the point of convergence of the psyche, between the different sides of the equator, in a downturn where the two pieces of the thalamus join.

- Structure: The organ is made from pinealocytes, the cells responsible for making melatonin, and glial cells, which proposition help and protection.
- Blood Supply: It gets a rich blood supply, second to the kidneys, allowing it to really convey synthetic substances into the circulatory framework.

Melatonin Production

Melatonin is the fundamental synthetic conveyed by the pineal organ, expecting a key part in coordinating rest wake cycles and circadian rhythms. The mix and appearance of melatonin are impacted by the light-faint cycle, with creation extending as a result of murkiness and reducing with light receptiveness.

- Light Sensitivity: The pineal organ's making of melatonin is coordinated by the suprachiasmatic center (SCN) of the operational hub, which gets light information from the eyes. This

information is given off to the pineal organ through the smart tactile framework.
- Circadian Rhythms: Melatonin synchronizes the body's internal clock with the external environment, propelling standard rest models and as a rule.

The Pineal Organ and Neural connections

Despite melatonin, the pineal organ is related with the creation and rule of a couple of neural connections, including serotonin and conceivably dimethyltryptamine (DMT), which are captured in mentality rule and changed states of comprehension.

Serotonin

Serotonin is a neural connection that expects a key part in mentality rule, yearning, and rest. It is in like manner a precursor to melatonin, making its presence in the pineal organ crucial for the blend of melatonin.

- Perspective Regulation: Raised levels of serotonin are connected with vibes of flourishing and fulfillment. The pineal organ's work in exchanging serotonin over totally to melatonin features its importance in both perspective rule and rest.

Dimethyltryptamine (DMT)

DMT is serious areas of strength for a compound regularly conveyed in humble amounts in the human body. A couple of experts guess that the pineal organ could convey DMT, particularly during uncommon experiences, for instance, close death states, significant consideration, or REM rest.

- Heavenly Experiences: DMT is known for provoking critical changes in acumen, including particular visual brain flights, changed sensation of the real world, and vibes of fortitude with the universe. These experiences eagerly seem to be depictions of supernatural feelings of excitement and otherworldly states in various traditions.

The Pineal Organ and Light Wisdom

The pineal organ's repugnance for light is maybe of its generally fascinating viewpoint, proposing a more significant relationship with our impression of this present reality and comprehension.

Photoreceptor Cells
The pineal organ contains photoreceptor cells like those found in the retina of the eyes. This similarity has driven a couple of researchers to suggest that the pineal organ created from an unrefined light-fragile organ, conceivably filling in as an internal "third eye."

- Groundbreaking Perspective: In lower vertebrates, for instance, reptiles and animals of land and water, the pineal organ capacities as a direct photoreceptor, affecting approaches to acting considering light receptiveness. In individuals, while the organ isn't directly photosensitive, it really answers light signals gave off through the eyes and the SCN.

Pineal Organ Calcification

Calcification of the pineal organ, a cycle where calcium stores assemble inside the organ, can incapacitate its capacity and is associated with various clinical issues.

Purposes behind Calcification

- Aging: Calcification of the pineal organ will in everyday augmentation with age, conceivably lessening its ability to make melatonin.
- Fluoride Exposure: A couple of examinations propose an association between's fluoride transparency and extended pineal organ calcification. Reducing fluoride confirmation could help with staying aware of the organ's prosperity.

- Diet and Lifestyle: Horrible eating routine and nonattendance of real work can add to calcification. A sound eating routine rich in cell fortifications and standard movement can help with diminishing calcification.

Effects of Calcification

- Rest Disorders: Diminished melatonin creation in view of calcification can provoke rest issues like lack of sleep.
- Perspective Disorders: Impaired serotonin rule could add to disposition issues, including gloom and strain.
- Significant Implications: As indicated by an extraordinary perspective, calcification is made sure to frustrate the pineal organ's ability to work with supernatural experiences and significant stimulating.

Decalcifying the Pineal Organ

Decalcifying the pineal organ remembers taking for lifestyle and dietary changes that help the organ's prosperity and ability.

Dietary Recommendations

- Normal Foods: Eating regular results of the dirt reduces receptiveness to pesticides and various fabricated materials that can add to calcification.
- Chlorophyll-rich Foods: Food sources like spirulina, chlorella, and wheatgrass help with detoxifying the body and sponsorship pineal organ prosperity.
- Turmeric and Ginger: These flavors have quieting properties and can maintain all around frontal cortex prosperity.

Lifestyle Practices

- Sunlight Exposure: Common receptiveness to ordinary light coordinates the pineal organ's advancement of melatonin and supports circadian rhythms.

- Hydration: Drinking a ton of water helps flush toxins from the body, supporting in everyday prosperity.
- Consideration and Mindfulness: These practices decline strain and advance loosening up, helping the pineal organ.

Supplements
- Iodine: Supports thyroid ability and may help with decreasing calcification.
- Magnesium: Key for by far most regularphysical processes, including loosening up and rest.
- Boron: Detoxifies the body and sponsorship by and large.

Getting over Science and Secret

The pineal organ fills in as a charming expansion between the spaces of science and powerful nature. While coherent assessment gives pieces of information into its physiological and biochemical abilities, old significant practices offer knowledge on its part in exciting higher states of awareness.

Sensible Exploration

- Research on Melatonin: Focuses on melatonin's part in overseeing rest and circadian rhythms highlight the pineal organ's significance in as a rule.
- Stimulating Research: Assessments concerning the effects of DMT and other hallucinogenics give significant pieces of information into the biochemical reason of captivated experiences.

Significant Practices

- Meditation: Old consideration practices line up with present day disclosures on the benefits of care and stress decline.
- Portrayal and Chanting: Systems, for instance, discernment and presenting have been used from here onward, indefinitely a truly prolonged stretch of time to start the pineal organ and work with supernatural stirring.

The pineal organ stays at the intermingling of science and power, encapsulating the mysteries of human mindfulness and the potential for tremendous change. Sorting out its development, capacity, and the biochemical cycles that help its powerful properties offers a careful perspective on its part in significant stimulating.

Billy Carson's examples, close by encounters from present day science, give a sweeping method for managing supporting and starting the pineal organ. As we continue to examine the practices and data including this baffling organ, we open new parts of care, defeating any issues between the physical and the supernatural areas.

In the going with parts, we will dive into practical purposes of pineal organ activation, researching how to harness its power for mindfulness, recovering, and change. Through these practices, you will open new components of care and partner with the more significant pieces of your being.

CHAPTER 7:

Practical Applications

Authorizing the pineal organ and opening its actual limit can incite huge mindfulness, repairing, and change. Integrating the data and deals with including the pineal organ into your ordinary presence can update your thriving and significant outing. In this part, we will research suitable usages of pineal organ establishment, focusing in on consideration, diet, lifestyle changes, and the use of advancement.

Reflection Techniques for Pineal Organ Order

Reflection is a mind boggling resource for empowering the pineal organ and progressing extraordinary exciting. A couple of unequivocal reflection techniques can work on the capacity of the pineal organ and work with additional significant states of comprehension.

Third Eye Meditation

Third eye examination revolves around the district between the eyebrows, related with the pineal organ and the Ajna Chakra. This preparing starts the organ and open the third eye.

Bit by bit directions to Practice Third Eye Meditation:

1. Find a Quiet Space: Pick a tranquil, pleasing place where you won't be disturbed.
2. Sit Comfortably: Sit in a pleasing circumstance with your spine straight and eyes shut.
3. Focus on Your Breath: Take significant, slow breaths to relax your body and mind.
4. Visualize the Third Eye: Guide your concentration toward the locale between your eyebrows. Imagine a splendid light or a glimmering circle here.
5. Chant or Repeat a Mantra: You can recount "Om" or any mantra that influences you to help with staying aware of focus.

6. Stay Present: Accepting that your cerebrum wanders, carefully return your thought to the third eye.

7. Practice Regularly: Plan to practice this reflection ordinarily for somewhere near 10-15 minutes.

Breathwork

Breathwork strategies, as Pranayama and the Breath of Fire, can enliven the pineal organ and sponsorship all around flourishing.

Breath of Fire:

1. Sit Comfortably: Find a pleasant arranged position with your spine straight.

2. Inhale Deeply: Take a full breath in through your nose.

3. Exhale Rapidly: Inhale out quickly through your nose while pulling your navel in towards your spine.

4. Continue Rhythmically: Continue with this quick, melodic unwinding for 1-3 minutes.

Insight Techniques

Insight incorporates using your imaginative brain to make mental pictures that help your targets and objectives. Envisioning the commencement of your pineal organ can update your appearance practice and work with supernatural stirring.

The best strategy to Practice Visualization:

1. Find a Quiet Space: Sit or rests in a pleasing position.
2. Focus on Your Breath: Take significant, slow breaths to relax your body and mind.
3. Visualize the Third Eye: Imagine an awe inspiring light or a sparkling circle at the region of your third eye.
4. Feel the Activation: Envision the light growing further and feel the energy in your third eye district.

5. Practice Regularly: Coordinate discernment into your ordinary reflection practice.

Dietary Practices for Pineal Organ Prosperity

The food you gobble up expects a gigantic part in the prosperity and capacity of your pineal organ. Certain food sources and upgrades can maintain the organ's capacity and help with diminishing calcification.

Food assortments to Help Pineal Organ Health:

1. Raw Cacao: Affluent in disease counteraction specialists and magnesium, rough cacao maintains all things considered prosperity.
2. Chlorophyll-rich Foods: Food sources like spirulina, chlorella, and wheatgrass help with detoxifying the body and support pineal organ prosperity.

3. Iodine-rich Foods: Sea development and iodine upgrades can help with decalcifying the pineal organ.
4. Turmeric: This zing has quieting properties and supports by and large prosperity.
5. Apple Juice Vinegar: Detoxifies the body and may assist with decalcification.

Supplements for Pineal Organ Activation:

1. Iodine: Supports the thyroid and may help with diminishing calcification.
2. Boron: Detoxifies the body and sponsorship overall prosperity.
3. Magnesium: Essential for by far most normalphysical cycles and supports loosening up and rest.
4. Melatonin: Upgrading with melatonin can maintain the ordinary capacity of the pineal organ.

Lifestyle Changes for Further developing Pineal Organ Capacity

Certain lifestyle changes can generally influence the prosperity and ability of your pineal organ. Coordinating these movements can overhaul your overall success and sponsorship your significant outing.

Diminish Fluoride Exposure

Fluoride is a common ally of pineal organ calcification. Diminishing receptiveness to fluoride can help with staying aware of the prosperity of your pineal organ.

Tips to Diminish Fluoride Exposure:

1. Drink Isolated Water: Use a water channel that wipes out fluoride.
2. Choose without fluoride Toothpaste: Settle on ordinary toothpaste without fluoride.
3. Avoid Dealt with Foods: Took care of food assortments every now and again contain

fluoride and other pernicious manufactured substances.

Sunshine Exposure

Ordinary sunshine expects a critical part in controlling the pineal organ's improvement of melatonin and supporting commonly circadian rhythms. Standard receptiveness to light, especially close to the start of the day, can help your pineal organ.

Tips for Light Exposure:

1. Spend Time Outdoors: Plan to spend something like 15-30 minutes outside each day.
2. Morning Sunlight: Morning light is particularly important for coordinating your circadian rhythm.
3. Limit Counterfeit Light: Diminishing receptiveness to fake light, especially blue light from screens, around evening time.

Rest Hygiene

Extraordinary rest tidiness maintains the customary ability of the pineal organ and the advancement of melatonin. Ensuring an anticipated rest plan and laying out a quiet rest environment can overhaul your overall success.

Tips for Better Rest Hygiene:

1. Maintain a Consistent Rest Schedule: Go to rest and stir at the same time reliably.
2. Create a Loosening up Environment: Make your room supportive for set down with open to bedding and unimportant light and disturbance.
3. Limit Stimulants: Avoid caffeine and electronic devices before rest time.

Involving Development for Pineal Organ Inception

Present day development offers various instruments and resources that can maintain the incitation of the pineal organ and redesign significant practices.

Binaural Beats and Isochronic Tones

Binaural beats and isochronic tones are sound advancements that can entrain brainwaves to express frequencies, progressing loosening up, reflection, and changed states of discernment.

Directions to Use Binaural Beats:

1. Choose the Right Frequency: Select binaural beats that emphasis on the best brainwave repeat (e.g., alpha, theta, or delta waves).
2. Use Headphones: Binaural beats require headphones for the sound effect on precisely work.
3. Relax and Listen: Find a pleasing spot to sit or rests, shut your eyes, and focus on the binaural beats for 15-30 minutes.

Light and Sound Machines

Light and sound machines use blends of light pulses and sound frequencies to invigorate the brain and advance significant loosening up and consideration.

Guidelines to Use Light and Sound Machines:

1. Follow Instructions: Each machine goes with unequivocal headings for use. Follow these warily to ensure safeguarded and practical use.
2. Create a Relaxing Environment: Use the machine in a quiet, pleasing space where you can loosen up without interferences.
3. Use Regularly: Incorporate the machine into your standard reflection or loosening up everyday practice for best results.

Organizing Practices into Everyday presence

Planning extraordinary practices into your everyday timetable can make solid areas for a for significant stimulating and the commencement of the pineal organ.

Ordinary Routine

Spreading out a dependable everyday timetable that integrates consideration, care, and other significant practices can maintain your trip towards significant stimulating.

Cautious Moments

Unite cautious minutes throughout the span of your day to remain related with your supernatural practice. This can consolidate short reflection gatherings, significant breathing, or simply stopping briefly to see the worth in the ongoing second.

Neighborhood Support

Attracting with a neighborhood comparable individuals can offer assistance and inspiration on your significant trip. Participate in consideration social affairs, significant retreats, or online conversations to connect with others who share your tendencies.

Starting the pineal organ and outfitting its actual limit can provoke huge personal development, recovering, and change. By uniting examination, dietary practices, lifestyle changes, and current advancement into your everyday daily schedule, you can maintain the prosperity and capacity of your pineal organ and work on your significant journey.

Billy Carson's examples, close by pieces of information from present day science and old knowledge, offer a broad framework for understanding and starting the pineal organ. As you continue to examine these practices and integrate them into your standard daily schedule, you will open new parts of care and set out on a remarkable journey toward significant stirring.

In the going with areas, we will dive into the singular experiences and accolades of individuals who have really authorized their pineal organs and explore the notable impact it has had on their lives. Through these records, you will secure further encounters and inspiration to continue with your own trip of self-disclosure and significant turn of events

CHAPTER 8:

Challenges and Misconceptions

While the excursion to actuate and bridle the force of the pineal organ can be groundbreaking, it is additionally loaded with difficulties and confusions. Understanding these difficulties and exposing normal fantasies is significant for a fruitful and legitimate investigation of the pineal organ. In this part, we will address the normal hindrances looked by people trying to enact their pineal organ, also as explain misinterpretations that can prevent progress.

Normal Difficulties in Pineal Organ Enactment

1. Conquering Skepticism

Wariness about the pineal organ's job in otherworldly arousing and cognizance can be a critical test. Many individuals view the pineal organ's otherworldly relationship as pseudoscience as opposed to authentic natural peculiarities.

Procedures to Address Skepticism:

- Education: Find out about the logical examination on the pineal organ and its capabilities. Understanding the biochemical cycles and physiological impacts can give a more adjusted viewpoint.
- Experience-Based Evidence: Participate in practices like reflection and breathwork, and notice the individual encounters and advantages firsthand.

2. Trouble in Contemplation and Focus

Contemplation is a critical practice for initiating the pineal organ, yet numerous people battle with keeping up with center or accomplishing a profound thoughtful state.

Procedures to Further develop Meditation:

- Begin Small: Start with more limited contemplation meetings and slowly increment the length as you become more agreeable.
- Utilize Directed Meditations: Directed reflections can assist you with remaining on track and give construction to your training.
- Make a Predictable Routine: Lay out a normal contemplation practice to construct discipline and further develop focus over the long run.

3. Tending to Pineal Organ Calcification

Calcification of the pineal organ is a typical issue that can impede its capability and lessen melatonin creation. Tending to calcification requires a diverse methodology.

Techniques to Battle Calcification:

- Diet and Nutrition: Consolidate food sources and enhancements that help decalcification and generally speaking pineal organ wellbeing.
- Way of life Changes: Diminish openness to fluoride, increment daylight openness, and keep up with great rest cleanliness.
- Customary Detoxification: Participate in detoxifying practices, for example, fasting or utilizing detoxifying spices to help generally speaking wellbeing.

4. Overseeing Expectations

A few people might have ridiculous assumptions regarding the results of pineal organ initiation, for example, anticipating quick or sensational otherworldly encounters.

Techniques for Overseeing Expectations:

- Set Sensible Goals: Comprehend that profound arousing and initiation of the pineal organ is a progressive interaction that requires some investment and reliable exertion.

- Zero in on the Journey: Underscore self-improvement and prosperity as opposed to exclusively zeroing in on unambiguous results or encounters.

Normal Misinterpretations About the Pineal Organ

1. The Pineal Organ is a "Mysterious" Organ

One normal confusion is that the pineal organ is simply otherworldly and not grounded in logical reality. While it has otherworldly importance, it likewise has irrefutable natural capabilities.

Clarification:

- Logical Basis: The pineal organ's job in creating melatonin and directing circadian rhythms is deeply grounded in logical examination. Its association with otherworldly encounters is an area of progressing investigation.

- Reconciliation of Science and Spirituality: Perceive that the pineal organ's mysterious affiliations supplement its logical comprehension, giving an all encompassing viewpoint on its capabilities.

2. The Pineal Organ Can Be Initiated Instantly

Certain individuals accept that the pineal organ can be in a flash enacted through unambiguous methods or substances. This confusion neglects the requirement for steady practice and way of life changes.

Clarification:

- Continuous Process: Actuation of the pineal organ and the related advantages require progressing practice, way of life changes, and tolerance. It is a slow cycle as opposed to a quick change.
- Comprehensive Approach: Viable initiation includes a blend of contemplation, dietary

changes, way of life alterations, and self-awareness.

3. Pineal Organ Actuation Equivalents Enlightenment

Another misguided judgment is that initiating the pineal organ naturally prompts illumination or otherworldly edification. While it can uphold profound development, it's anything but an assurance of accomplishing illumination.

Clarification:

- Individual Growth: Initiation of the pineal organ can upgrade individual mindfulness and profound practices, however edification is a more extensive excursion including self-revelation, inward work, and reconciliation of different practices.
- Continuous Practice: Profound development and illumination require proceeded with exertion and investigation past the actuation of the pineal organ.

4. Just Certain Practices Are Effective

A few people trust that main explicit practices, like specific contemplation methods or enhancements, are compelling for enacting the pineal organ. This can restrict investigation and forestall the coordination of different practices.

Clarification:

- Changed Approaches: Various practices and approaches can be successful in enacting the pineal organ. It's vital to investigate a scope of strategies and find what turns out best for you by and by.
- Personalization: Designer your practices to your exceptional necessities and inclinations, consolidating different strategies like contemplation, perception, and way of life changes.

Tending to Difficulties and Confusions Practically speaking

1. Constant Learning

Remain informed about the most recent exploration and practices connected with the pineal organ. Drawing in with refreshed information and various viewpoints can upgrade your comprehension and viability in enacting the pineal organ.

2. Looking for Guidance

Consider looking for direction from experienced experts or educators who can offer help, bits of knowledge, and customized counsel on your excursion. Drawing in with a local area of similar people can likewise offer support and inspiration.

3. Embracing Tolerance and Persistence

Perceive that the excursion to enact the pineal organ and accomplish otherworldly development is a nonstop cycle. Embrace tolerance, steadiness, and self-empathy as you explore difficulties and work toward your objectives.

Understanding and tending to the difficulties and misguided judgments encompassing the pineal organ is vital for a compelling and genuine investigation of its true capacity. By perceiving normal impediments, exposing fantasies, and taking on an all encompassing methodology, you can explore your excursion with more noteworthy lucidity and certainty.

Billy Carson's lessons and current logical experiences give significant structures to understanding and defeating these difficulties. As you proceed to investigate and incorporate practices for pineal organ enactment, stay open to learning, look for direction, and embrace the excursion with persistence and commitment.

In the accompanying parts, we will investigate individual tributes and examples of overcoming adversity of people who have explored these difficulties and experienced groundbreaking outcomes through pineal organ enactment. Their accounts will offer motivation and down to earth experiences for your own excursion toward otherworldly arousing and self-revelation.

CONCLUSION:

The Future of Pineal Gland Research

As we close our investigation of the pineal organ, its actuation, and its suggestions for individual and otherworldly development, it is fundamental for anticipate the eventual fate of pineal organ research. Progresses in science, innovation, and profound practices are ceaselessly extending how we might interpret this cryptic organ and its job in human awareness.

Arising Patterns and Future Bearings

1. Propels in Neuroscience

Neuroscience research is progressively centered around figuring out the pineal organ's job in the mind's general working. Future investigations might reveal more about how the pineal organ communicates with other cerebrum districts and impacts cognizance and comprehension.

- Neuroimaging Techniques: Advances in neuroimaging, like utilitarian X-ray and PET sweeps, could give further bits of knowledge into the pineal organ's movement and its association with different mental states and encounters.

- Hereditary Research: Exploring the hereditary variables affecting pineal organ capability might uncover new data about its part in wellbeing and illness.

2. Investigation of Hallucinogenic Research

The investigation of hallucinogenics, especially DMT, which is conjectured to be created by the pineal organ, is building up some decent forward movement. Examination into the impacts of hallucinogenics on awareness and otherworldly encounters might additionally enlighten the pineal organ's true capacity.

- Clinical Trials: Continuous clinical preliminaries investigating the helpful utilization of hallucinogenics could give significant bits of knowledge into their effect on the pineal organ and otherworldly encounters.
- Wellbeing and Efficacy: Future exploration should address security and viability concerns, guaranteeing that the utilization of

hallucinogenics for self-awareness and mending is led mindfully.

3. Combination of Innovation and Spirituality

Mechanical headways are progressively being coordinated into otherworldly works on, including those pointed toward enacting the pineal organ. Future advancements might upgrade our capacity to investigate and figure out the pineal organ's true capacity.

- Biofeedback Devices: Arising biofeedback advancements could offer constant experiences into physiological changes related with pineal organ enactment, considering more designated rehearses.
- Computer generated Reality (VR): VR encounters intended for contemplation and profound investigation might give vivid conditions that help pineal organ actuation and self-improvement.

4. Comprehensive Methodologies and Personalization

The fate of pineal organ examination will probably underscore customized and comprehensive methodologies, perceiving the interesting necessities and encounters of people.

- Tweaked Protocols: Exploration might prompt customized conventions for pineal organ initiation, custom-made to individual wellbeing profiles, otherworldly objectives, and way of life factors.
- Interdisciplinary Collaboration: Coordinated effort between researchers, profound specialists, and all encompassing wellbeing specialists will encourage a more exhaustive comprehension of the pineal organ's true capacity.

Suggestions for Individual and Otherworldly Development

As how we might interpret the pineal organ keeps on advancing, it will have significant ramifications for individual and profound development. The joining of logical experiences with profound practices can offer new open doors for self-revelation and change.

1. Improved Self-Awareness

Expanded information about the pineal organ's capability and its job in cognizance can prompt more prominent mindfulness and individual knowledge. By understanding how to help and initiate the pineal organ, people can develop a more profound association with themselves and their otherworldly desires.

2. Further developed Wellbeing and Well-Being

Examination into the pineal organ's part in managing rest, temperament, and generally wellbeing can prompt superior prosperity and personal satisfaction. Incorporating rehearses that help the pineal organ can upgrade actual wellbeing, profound equilibrium, and otherworldly satisfaction.

3. Extended Profound Practices

As new bits of knowledge and innovations arise, people can investigate a more extensive scope of profound practices and instruments intended to help pineal organ enactment and otherworldly development. This extended scope of practices can advance one's profound excursion and give new pathways to investigation and self-revelation.

Source of inspiration

As we plan ahead, it is essential to stay open to new revelations and progressions in pineal organ research. Whether through logical request, profound investigation, or individual practice, every individual has the chance to add to and benefit from this advancing field.

- Take part in Research: Remain informed about the most recent improvements in pineal organ

research and investigate new discoveries that might upgrade your comprehension and practice.
- Practice Mindfulness: Keep on integrating rehearses that help pineal organ initiation and self-improvement, like contemplation, sound living, and profound investigation.
- Share Knowledge: Offer your encounters and experiences with others, adding to an aggregate comprehension of the pineal organ's true capacity and its effect on human cognizance.

Last Considerations

The excursion to comprehend and initiate the pineal organ is an entrancing and developing investigation of the convergence among science and otherworldliness. As examination propels and new revelations are made, we will keep on uncovering the secrets of this noteworthy organ and its job in our lives.

Billy Carson's lessons, joined with progressing logical exploration, give a rich establishment to investigating the pineal organ's true capacity. By embracing both logical request and otherworldly practice, we can extend how we might interpret the pineal organ and its groundbreaking prospects.

As we push ahead, let us approach this excursion with interest, transparency, and a guarantee to individual and aggregate development. The eventual fate of pineal organ research holds invigorating opportunities for improving our mindfulness, prosperity, and profound satisfaction

APPENDIX

The addendum gives extra assets, references, and useful apparatuses to help perusers in their investigation of the pineal organ and its actuation. It incorporates further understanding materials, suggested rehearses, and pertinent examinations to upgrade understanding and use of the ideas talked about in the book.

Extra Assets

1. Books and Publications

- "The Pineal Organ: The Eye of God" by Dr. John M. McCleary
An inside and out investigation of the pineal organ's set of experiences, capabilities, and profound importance.

- "Arousing the Pineal Organ: A Complete Manual for Third Eye Initiation" by Billy Carson
A point by point guide on procedures and practices for enacting the pineal organ, wrote by Billy Carson.

- "The Soul Atom: Terence McKenna's Investigation of the Pineal Organ and DMT" by Rick Strassman
This book explores the job of DMT in the pineal organ and its suggestions for awareness and otherworldliness.

2. Logical Diaries and Articles

- "Melatonin and the Pineal Organ: A Review"
This audit article investigates the job of melatonin in managing circadian rhythms and its effect on by and large wellbeing.

- "Calcification of the Pineal Organ: A Review"
An outline of the elements adding to pineal organ calcification and possible intercessions for decalcification.

- "Hallucinogenics and the Pineal Organ: A Survey of the Evidence"
A survey of examination on hallucinogenics, especially DMT, and their consequences for the pineal organ and cognizance.

3. Reflection and Care Resources

- "Headspace"
An application offering directed reflections and care practices to help mental prosperity and concentration.

- "Calm"
An application giving assets to contemplation, unwinding, and rest improvement.

- "Knowledge Timer"
A contemplation application highlighting a great many directed reflections, including those zeroed in on pineal organ enactment.

4. Dietary and Way of life Resources

- "The Detox Book: How to Detoxify Your Body to Eliminate Destructive Poisons" by Dr. T. J. Clark
A manual for detoxification rehearses that help generally speaking wellbeing and pineal organ capability.

- "The Whole30: The 30-Day Manual for Complete Wellbeing and Food Opportunity" by Melissa Hartwig Urban

A nourishing aide zeroing in on entire food varieties and dietary changes that can help pineal organ wellbeing.

5. Innovation and Tools

- "Muse: The Cerebrum Detecting Headband"
A neurotechnology gadget that gives constant criticism on reflection practice and brainwave movement.

- "Apollo Neuro: Wearable Innovation for Stress Relief"
A wearable gadget that utilizations delicate vibrations to help unwinding and stress decrease.

- "Light and Sound Machines"
Gadgets intended to animate the cerebrum with light heartbeats and sound frequencies to improve reflection and unwinding.

Suggested Practices and Procedures

1. Pineal Organ Reflection Protocol

1. Preparation: Track down a peaceful and agreeable space. Sit with your spine straight and loosen up your body.
2. Breathing: Take profound, slow breaths to quiet your brain and body.
3. Visualization: Spotlight on the region between your eyebrows. Envision a shining light or circle around here.
4. Affirmations: Use certifications, for example, "I enact my third eye and welcome further mindfulness" to support your goal.
5. Duration: Practice for 10-15 minutes day to day.

2. Detoxification and Dietary Adjustments

1. Avoid Fluoride: Use sans fluoride toothpaste and separated water to diminish fluoride consumption.
2. Incorporate Iodine: Incorporate iodine-rich food sources like ocean growth and iodine enhancements to help pineal organ wellbeing.
3. Consume Antioxidants: Eat food varieties wealthy in cell reinforcements, like berries, nuts,

and green verdant vegetables, to help detoxification.

3. Daylight Openness and Rest Hygiene

1. Morning Sunlight: Spend somewhere around 15-30 minutes outside in normal daylight every morning to manage circadian rhythms.
2. Limit Blue Light: Lessen openness to electronic screens and counterfeit lighting at night to help regular melatonin creation.
3. Consistent Rest Schedule: Keep a customary rest timetable to upgrade generally wellbeing and backing pineal organ capability.

Pertinent Examinations and Papers

- "The Job of Melatonin in Circadian Beat Regulation"
A review investigating melatonin's effect on circadian rhythms and rest designs.

- "Fluoride and Pineal Organ Calcification: An Orderly Review"

Research on the connection between fluoride openness and pineal organ calcification.

- "Impacts of Contemplation on Pineal Organ Function"
A review looking at the effect of reflection rehearses on the pineal organ and by and large prosperity.

Glossary of Terms

Pineal Gland: A little, pea-formed endocrine organ situated in the mind that produces melatonin and directs rest wake cycles.
- Melatonin: A chemical created by the pineal organ that controls rest and circadian rhythms.
- Calcification: The collection of calcium salts in tissues, which can prompt weakened capability of the pineal organ.
- DMT (Dimethyltryptamine): A strong hallucinogenic compound that is conjectured to be created by the pineal organ and related with changed conditions of cognizance.

- Third Eye: A term used to depict the pineal organ's job in profound mindfulness and instinct.

This informative supplement fills in as a complete asset to help your proceeded with investigation of the pineal organ. By using these materials and practices, you can extend your comprehension and improve your excursion toward initiation and profound development.

ACKNOWLEDGEMENTS

Composing this book has been a significant excursion, and it is with profound appreciation that I recognize the commitments and backing of the people who made this work conceivable.

To Billy Carson: Your pivotal bits of knowledge and devotion to investigating the secrets of the pineal organ have been a monstrous wellspring of motivation. Your work has enlightened the way for the majority as well as given a vigorous system to understanding and initiating the pineal organ. Much thanks to you for your enthusiasm and obligation to this field.

To Specialists and Scientists: The exploration and studies referred to in this book have been instrumental in giving a logical premise to the conversation on the pineal organ. I stretch out my appreciation to every one of the scientists whose work has added to how we might interpret this entrancing organ and its part in human wellbeing and awareness.

To the Specialists and Experts: Your direction and viable experiences into reflection, way of life practices, and otherworldly development have incredibly enhanced the substance of this book. Your commitments have given significant viewpoints on enacting the pineal organ and its applications in self-awareness.

To My Family and Friends: Your resolute help, support, and understanding have been critical all through this creative cycle. Much thanks to you for your understanding and for trusting in the significance of this work.

To My Readers: Your interest and obligation to investigating the secrets of the pineal organ motivate me. I trust this book gives significant bits of knowledge and reasonable direction on your excursion toward actuation and otherworldly development.

To the Numerous Anonymous Contributors: This book remains on the shoulders of endless people whose aggregate information and encounters have formed how we might interpret the pineal organ. Your commitments, whether through distributed work or individual practice, are profoundly valued.

Much obliged to you for your commitments, backing, and motivation. This work is a demonstration of the cooperative exertion and shared enthusiasm for revealing the capability of the pineal organ and its effect on human cognizance and prosperity.

www.ingramcontent.com/pod-product-compliance
Lightning Source LLC
Chambersburg PA
CBHW071938210526
45479CB00002B/726